SCIENCE MUSEUM

Leonardo da Vinci's Aeronautics

by

CHARLES H. GIBBS-SMITH

'Nature cannot again produce his like.'

– *Melzi, on Leonardo* (1519)

GW00695768

LONDON

HER MAJESTY'S STATIONERY OFFICE

1967

Fig. 1. Self-portrait of Leonardo da Vinci.

LEONARDO DA VINCI
(1452—1519)

Perhaps the greatest universal genius of history, Leonardo da Vinci (1452–1519) provides an absorbing chapter in aeronautical history. He was the first man of high scientific attainments to investigate the problems of flight, and although a few of his aeronautical notes and drawings may have been lost, the bulk seem to have survived, and it is fairly clear where his thoughts and feelings lay. Most of his work dealt with flapping-wing aircraft (ornithopters), and his concern with the human imitation of birds and bats amounted to an obsession. As a result of his emotional rather than rational approach to flying, he did not on the whole subject it to the disciplined scrutiny he applied to other of his scientific activities, although his aeronautical work was shot through with sparks of brilliant and prophetic ingenuity. Through no fault of his own, Leonardo had no direct influence upon the development of flying, for the world knew nothing of his aeronautical work until late in the nineteenth century, when both practical and theoretical aeronautics were bringing near the final achievement of human flight (see below). One must therefore view Leonardo in what is virtually a historical vacuum.

He left a large body of manuscripts—amounting to something above 35,000 words and over 500 sketches—dealing with flying machines, the nature of the air, and bird flight. This corpus also includes one small complete work on bird flight, *Sul Volo degli Uccelli* (dated 1505) in the Turin Library. The rest of the material consists of a mass of scattered notes and sketches, most of them contained in Manuscripts A to M in the Library of the Institut de France in Paris, and in the great *Codice Atlantico* in the Ambrosiana Library at Milan.

Contrary to popular belief, Leonardo did not base most of his aeronautical speculations and designs on a careful study of bird and bat flight, but on comparatively superficial observations; for his detailed study of birds came after, not before, the famous designs for ornithopters; and there is evidence that, as he learnt more of birds in his old age, he was moving away from his belief in full flapping flight. As he was a careful observer of the gliding and soaring flight of birds, it is surprising that he never considered a proper controlled glider in the modern sense, although the last drawing he made suggests he might have done so later. The glider proper was a type of machine he might well have pioneered, had his peculiar psychological constitution not driven him first to his obsessional concern with human ornithopter flight. As already said, a factor beyond the passion for truth and enquiry was at work, a factor which indeed hindered and twisted his investigations; for, inextricably interwoven with his desire to impartially investigate the problems of flight, was his powerful symbolic interest in the romantic idea of flight.

3

His curious emotional make-up in other respects has been exhaustively studied by historians and psychologists; but the deep-seated urge towards power, escape and freedom, which is clearly expressed in both his words and his behaviour where flying was concerned, has not been paid the attention it deserves. Typical of this was his habit of buying caged birds in the market place at Florence, only to release them, and have the satisfaction of seeing them fly off to freedom. One biographer of Leonardo – Vallentin – has referred to his concern with flight as "the most obsessing, most tyrannical of his dreams"; and when one reads the crowning passage of his treatise on bird flight, the force of his feelings is most evident:

> "The great bird will make its first flight upon the back of the great swan [a reference to Mount Ceceri (Swan) near Fiesole] filling the whole world with amazement, and all records with its fame; and it will bring eternal glory to the nest in which it was born".

That is the cry of a poet, not a scientist.

So it is the powerful emotion and symbolism of flight, as well as his deep scientific curiosity, which we encounter immediately when approaching Leonardo's manuscripts, and which overrode his scientific judgment, and prevented this formidable man from advancing as far in aviation as he did in other branches of science.

There were also two vital misconceptions which he allowed to possess him in his early work, which retarded his whole aeronautical development, so that he was only reaching towards more valid conclusions in his old age; these he could easily have achieved in his early and middle life. Nevertheless, we find the utmost ingenuity in his designs, and in the constant play of his genius.

The first of Leonardo's basic misconceptions was that man has the necessary muscle-power and skill to emulate the birds, "You will perhaps say," he writes, "that the sinews and muscles of a bird are incomparably more powerful than those of man . . . but the reply to this is that such great strength gives it a reserve of power beyond what it ordinarily uses to support itself on its wings; since it is necessary for it, whenever it may so desire, either to double or treble its rate of speed in order to escape from its pursuer, or to follow its prey." Man has, therefore (he argues), adequate power for flight without such a surplus. But Leonardo, who dissected both birds and men, was deliberately misleading himself at the command of his inner tyranny. For man has but a small fraction of muscle power to work his arms and legs—some 20–25% of his total weight—compared with the bird's flying muscles, which can weigh up to nearly half its total weight, to say nothing of the vital differences in metabolism and other factors. With modern ultra-lightweight materials, aerofoil design, propeller design, and transmission techniques, it has proved just possible to build and to fly man-powered aeroplanes for short distances. But at any time before this century, it was quite impossible; in Leonardo's day it was a romantically hopeless idea, as Borelli was to demonstrate convincingly in his work *De motu animalium*, published in 1680.

4

During the later years of his life, Leonardo's fantasies began slacking in their pressure, and his profound good sense took over at least to the extent of seeing man-powered flight as limited to a more rational machine, which incorporated fixed-wing surfaces. But his early—and chief—designs for flying machines remain a charmingly outrageous monument to his brilliant ingenuity, as will be seen in his drawings and descriptions. In none other of his scientific pursuits, where controlled curiosity held sway, would he have been guilty of such schemes.

Leonardo's second fundamental misconception was typical of his time, and indeed persisted until Sir George Cayley, in 1808, realised what actually happens to a bird's wings in flight. It was that birds sustain and propel themselves through the air by beating their wings downwards and backwards. "Write of swimming under water," Leonardo says, "and you will have the flight of the bird through the air." And when speaking of one of his aircraft, he says "the wings have to row downwards and backwards in order to keep the machine up, and so that it may progress forward." A bird does not, and cannot, beat backwards on the downstroke; such behaviour would virtually capsize the creature at every wing-beat, and it would proceed through the air rather like a bucking bronco.

Most birds possess four or five outer primary feathers, each with little or no vane in front of the shaft, and a large vane behind. When beaten downwards, these feathers twist into miniature high-speed propellers and pull the bird forwards, whilst the inner wing provides most of the lift. Some birds do not have these multiple 'propellers'; with various seabirds, as well as some large land birds with low aspect-ratio wings, the whole wing-tip assembly of feathers twists collectively into one large propeller on the downstroke; this is also the case with bats. On the upstroke, there is a similar wing action in the opposite direction, but reduced in effect, since the twisting effect is being exerted against the natural camber of the wing.

Leonardo had a profound knowledge of machines, of mechanical essentials, and he was certainly studying the structure of birds' wings for many years; he also understood the passive action of the airscrew (see later); so it is a little surprising that he did not arrive at a true appreciation of the main factors involved in bird propulsion.

Sir Kenneth Clark has pointed out the "curious fact that these records [*i.e.* his notebooks and manuscripts] only begin when Leonardo was thirty, an age when the average busy man ceases to take notes." As a result, we cannot know precisely when he started thinking, either inquisitively or compulsively, about the problem of human aviation. But the surviving evidence points to the most productive period of his aeronautical activity as lying within the period of his service with Ludovico Sforza, Duke of Milan, which extended from about 1482 to 1499, with the years 1486 to 1490 being particularly significant in Leonardo's sphere of aviation. During this period, he seems also to have been continually occupied with numerous varieties of engineering, which included armaments and fortifications, and his ingenuity was at its height.

Over the whole of the span of his aeronautical activities, Leonardo's aircraft may be divided into nine main types—the approximate dates are given in brackets—in which the word ornithopter is taken to mean a man-powered ornithopter, unless otherwise stated:*

TYPE A.   Prone ornithopter (1485–87); Figs. 2–8;
TYPE B.   Boat-shaped ornithopter (*c.* 1487); Figs. 9, 10;
TYPE C.   Standing ornithopter (1486–90); Figs. 11, 12;
TYPE D.   Powered ornithopter (1495–97); Fig. 13;
TYPE E.   Semi-ornithopter (1497–1500); Fig. 14;
TYPE F.   Falling-leaf glider (1510–15); Fig. 15;

. . . . .

TYPE G.   Finned Projectile (*c.* 1485); Fig. 23;
TYPE H.   Parachute (1483–86); Fig. 24;
TYPE I.   Helicopter (1486–90); Fig. 25.

The machines are here dealt with in this order; but after Type F will come a brief discussion of Leonardo's flap-valve wings, wing-construction, flight-control, and wing-testing rigs (Figs. 16–22), then notes on his finned projectiles, parachute, helicopter, and inclinometer (Figs. 23–26), followed by notes on his work on bird-flight, and other matters.

---

*This new classification differs from that which I suggested in the first edition of my Science Museum handbook *The Aeroplane; an Historical Survey* (1960), and is a result of my having now re-worked the whole subject of Leonardo's aeronautics, and also having taken advantage of Carlo Pedretti's dating of the *Codice Atlantico.*

Type A. Prone Ornithopters (1485-90)

The best-known aircraft designed by Leonardo are certainly his man-powered ornithopters; and we may denote his prone type of machine as Type A, where the pilot is either secured in a frame—which was his earliest conception—or lies on a board, and secured by hoops, beneath a framework supporting the fulcrum or hinges of his system of beating wing-spars.

The main source of power for effecting the downstroke of the wings is always one or both of the legs and feet; his earliest design (Figs. 2, 3) shows both the feet in stirrups operating the downstroke, while the slightly later ones (Figs. 4–7) show only one leg and foot operating the downstroke by a stirrup; in Fig. 6 this force is augmented by hand-worked cranks. The transmission always includes a multiple pulley system.

The recovery upstroke is shown at first (Figs. 2, 3) being solely undertaken by the hands, operating two levers, one of which is clearly seen hanging down in Fig. 2 and looking like an upturned crutch, which movement also twists the wings back to their normal 'resting' position (see below). He also suggested a spring to bring about the recovery stroke, and also the pulling up of both feet in stirrups simultaneously.

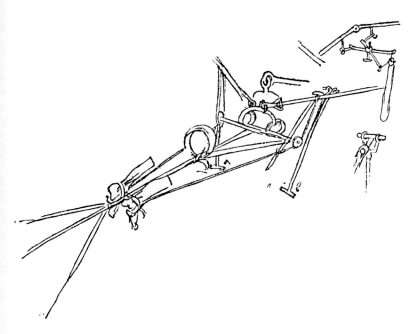

Fig. 2. Type A, prone ornithopter, with both legs moving together, and wings hand-operated on the up-stroke: c. 1487.

Fig. 3. The same type of machine as shown in Fig. 1, sketched as if in flight:
c. 1487.

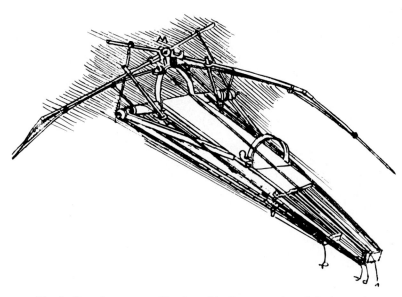

Fig. 4. Type A, prone ornithopter, with alternate action of the legs to lower
and raise the wings: 1486–90.

In most of Leonardo's developed ideas for ornithopter flight, each of his wings comprises the structure of a main spar, to which are added one or more articulated 'fingers', which hinge downwards from the end of the spar. In order to get the 'rowing' action of the wings—downwards and backwards—which he thought birds employed for propulsion, Leonardo generally arranged for the wing-spar and finger(s) to be collectively twisted, so that the down-beating fingers—operated by cables running out from pulleys on the central supporting member—in fact beat both downwards and backwards, thus simultaneously producing a lift component and a thrust component. He also provided for a bird-type tail-unit, and went on from there to a cruciform type.

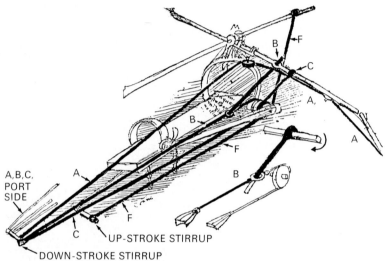

Fig. 5. Type A, prone ornithopter, annotated and drawn over to show the cable runs: *c.* 1487.

Leonardo's typical wing-operation therefore involved three simultaneous movements, *i.e.* (a) beating down; (b) twisting; and (c) the added beating down of the finger(s), which—by virtue of the twisting—beat backwards as well. I have included one of Leonardo's drawings (Fig. 5), which I have drawn-over and annotated. Here the stirrup effecting the downstroke sim-ultaneously involves these three movements: cable C pulls down the spar; B twists the spar (Leonardo includes a separate sketch of this system, just below the main drawing); and A, *via* a pulley on the support, runs out to the finger, and pulls it down, which—owing to the twisting movement of B—beats both downwards and backwards. The stirrup in this drawing which effects the upstroke, operates the cables FF, which pull down the extensions of the two spars beyond their fulcrum, and hence raise the two wings, whilst twisting back the spars to their normal 'resting' position.

Speaking about one of these prone-type machines, Leonardo writes: "This machine should be tried over a lake, and you should carry a long wineskin as a girdle, so that in case you fall, you will not be drowned".

But by far the most remarkable feature of any of the prone designs is seen in Fig. 6, where the machine is equipped with a sophisticated type of flight-control—the first in history—by means of a head-harness operating a cruciform elevator-cum-rudder: this is shown twice, once in place on the aircraft where the device is seen down to the right, impinging on the edge of the paper (with its head-harness out in front); and again, attached to the pilot's head, in a separate drawing in the top right hand corner, which is shown enlarged in

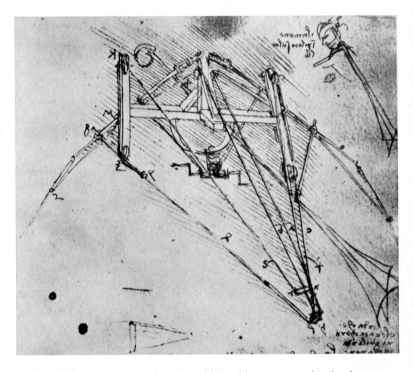

Fig. 6. Type A, prone ornithopter, with head-harness operating the elevator-cum-rudder: 1486–90.

Fig. 6a. This cruciform control-unit was not to be re-introduced until Sir George Cayley used it in 1799. What is so strange is that in no other surviving drawings by Leonardo does this device appear again. This machine has the added interest of the two hand-operated cranks below the supporting frame, to assist in the down-beat of the wings.

(*Right*)
Fig. 6a. Close-up of the head-harness and tail-unit in Fig. 6.

(*Below*)
Fig. 7. Type A, prone ornithopter, with alternate action of the legs, and four wings: 1486–90.

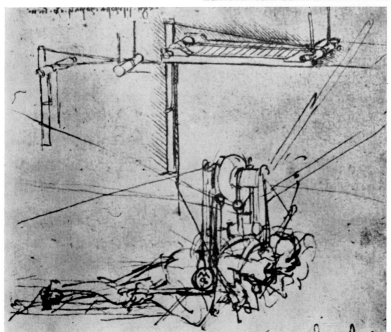

11

The rough but vivid sketch shown in Fig. 7 is of an ingenious machine with four wings, which beat with one up and one down on either side. Leonardo uses here one of his favourite transmission systems, with cables running over drums: in this case, two of the wing-spars are attached to a large rotating disc, aft of the top drum, which are caused to rock from side to side, with one wing being beaten down while the other is simultaneously pulled up: the other pair of wings rock on fulcrums, and the ends of their levers are attached by joints to the cables which run up vertically to operate the upper drum and disc, thus producing beats of the wings in opposite directions to those rocking with the disc. Beside this drawing Leonardo has written: "This can be made with one pair of wings, and also with two. If you wish to make it with one, the arms will raise it by means of a windlass, and two vigorous kicks with the heels will lower it, and this will be useful. And if you wish to make it with two pairs, when one leg is extended it will lower one pair of wings, and at the same time the windlass worked by the hands will raise the others, helping also considerably those that fall; and by turning the hands first to the right, and then to the left, you will help first the one, then the other."

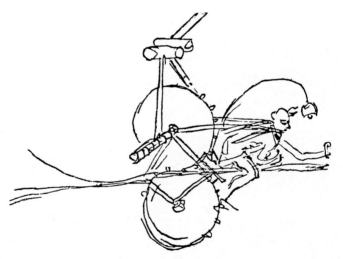

Fig. 8. Sketch of a semi-prone ornithopter, showing gear train, and cog-pedal transmission: c. 1485.

12

Away to one side of Leonardo's main stream of activity, and dating from about 1485, the same period as the early Type A designs, is a wonderfully wild little sketch—one might almost say a desperate sketch—of a kind of semi-prone machine, to which Leonardo never returns (Fig. 8): the pilot, oddly reminiscent of a racing cyclist, operates a gear-train transmission by pedalling direct on the cogs of the lower wheel, while simultaneously seeming to operate hand levers as well. The bell-like device attached to an outrigger, which is seen in front of the pilot's head, is clearly an inclinometer, as also seen, in a slightly different form, in Fig. 26 on page 31.

Type B.   Boat-Shaped Ornithopters (c. 1487)

Type B is an ornithopter with a boat-shaped fuselage or 'car', of which there are only two fairly explicit illustrations (Figs. 9, 10), one of which clearly implies a similar car, but with a different transmission system (Fig. 10). Leonardo has now arrived at the concept of a vehicle in which the pilot sits or stands, and in which he is at one remove, so to speak, from the actual power transmission. In Fig. 9 the pilot operates the wings *via* a quick thread jack, with a fulcrum in the form of a roller fixed to each gunwale of the car, under which the rowing-spars pass. It is somewhat hard to see how Leonardo here intends to effect the thrust component of the wings. But he must have known that no device such as he here suggests could possibly beat the wings fast enough for sustentation and propulsion. A notable feature of this design, by the way, is the properly large spreading tailplane (cum-elevator?), which is closely similar to that suggested by Henson in 1843. The second machine (Fig. 10)—which also clearly implies a pilot in a 'car'—shows one of Leonardo's favourite transmission devices, with cranks operating rollers, over which the

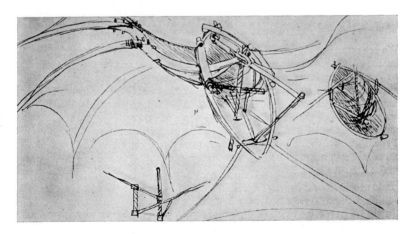

Fig. 9. Type B, boat-shaped ornithopter, with quick thread jack transmission: *c.* 1487.

13

cables pass to the wings. There is also here no hint of wing-twisting to effect the thrust component. Both these designs were, I feel, just general ideas thrown off without any intention, at that stage, to work out complete propulsion systems.

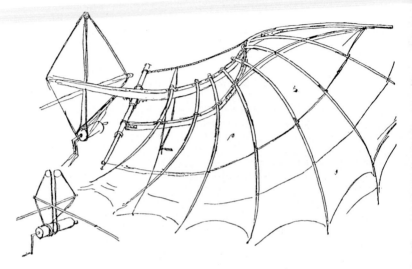

Fig. 10. Type B, boat-shaped ornithopter, with crank and pulley transmission: *c*. 1487.

Type C.   Standing Ornithopters (1486-90)

We are now reaching the phase in which Leonardo's sense of phantasy and his passion for ingenuity, are gaining ground. The date would be toward 1490, and he is obviously becoming oblivious of practicability, The prone ornithopters, although no human being could possibly sustain either the necessary speed of flapping by means of those cables, or the effort necessary for continued flight—quite apart from the basically wrong principle of operation—were nevertheless logically conceived and executed, and had a semblance of feasibility. With the Type B, and its quick thread jack transmission, we begin to arrive at the threshold of phantasy.

We will designate as Type C the standing ornithopter, in which the flashes of genius to be found in them diminish neither in quantity nor quality; but the vehicle Leonardo now proposes is nothing short of a chimera. In Fig. 11 is shown his new conception of flight; a man standing in a bowl-shaped aircraft, operating four beating wings *via* a massive transmission of hand-and-foot-operated drums, treadles and the like, at whose weight the aerodynamicists's mind could only boggle. The obsessional pursuit of ingenuity is also now in

14

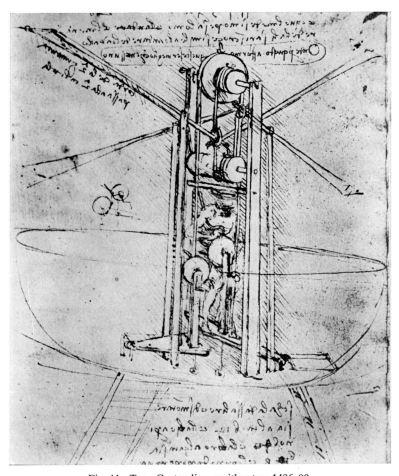

Fig. 11. Type C, standing ornithopter: 1486–90.

full flood, as scores of Leonardo's drawings testify; they show his pulley-and-drum transmission in every conceivable variation. This ingenuity also extends to such devices as the retractable undercarriage and entry ladder for a Type C standing ornithopter—shown in Fig. 12—which would of course send up the weight even further, especially as a separate windlass is provided to effect the retraction.

Besides some sketches of the transmission system, Leonardo wrote these words about the Type C in general: "I conclude that the upright position is more useful than face downwards, because the instrument cannot get overturned, and on the other hand the habit of long custom requires this. And the raising and lowering will proceed from the lowering and raising of the two

15

legs, and this is of great strength, and the hands remain free; whereas if it were face downwards, it would be very difficult for the legs to maintain themselves in the fastenings of the thighs."

Along with the general sketch of this monstrous type of craft (Fig. 11) Leonardo says: "This man exerts with his head* a force that is equal to two hundred pounds, and with his hands a force of two hundred pounds, and this is what the man weighs. The movement of the wings will be crosswise, after the manner of the gait of a horse. So for this reason I maintain that this

---

*The reference to his head is figurative, and includes the muscles of the neck and chest. The head seems to press against a pad of some kind beneath the lower drum, where a cross-member is fixed.

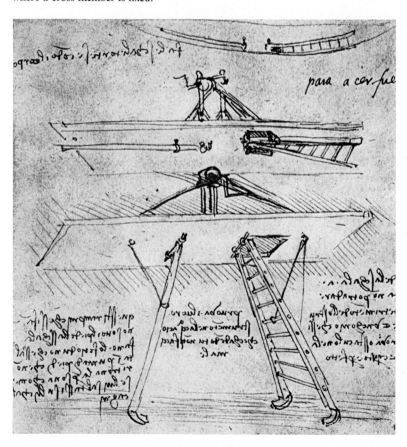

Fig. 12. Retractable undercarriage mechanism of the Type C, standing ornithopter: 1486–90.

method is better than any other. Ladder for ascending and descending; let it be twelve braccia high, and let the span of the wings be forty braccia, and their elevation eight braccia, and the body from stern to prow twenty braccia, and its height five braccia, and let the outside cover be all of cane and cloth." And beside his detailed drawings of the retractable undercarriage (Fig. 12) are these words: "Make the ladders curved to correspond with the body [*i.e.* bowl]. When the foot of the ladder *a* touches the ground, it cannot give a blow to cause injury to the instrument, because it is a cone which buries itself, and does not find any obstacle at its point, and this is perfect. Make trial of the actual machine over the water so that if you fall, you do not do yourself any harm. These hooks that are underneath the feet of the ladder act in the same way as when one jumps on the points of one's toes, for then one is not stunned, as is the person who jumps upon his heels. This is the procedure when you wish to rise from an open plain: these ladders serve the same purpose as the legs, and you can beat the wings whilst rising. . . . But when you have raised yourself, draw up the ladders as I show in the second figure above." It is interesting to note that in the topmost sketch, Leonardo has shown how the leg and ladder, which constitute the undercarriage, are to be neatly tucked up within the contour of the bowl-fuselage. This is not evidence that he was particularly aware of the importance of aerial streamlining; most of his attention in this subject was paid to marine streamlining.

The business of flying has now got completely out of hand in Leonardo's mind—apart from this machine having the first retractable undercarriage in history—it is best simply to record it, and lay it to rest. Perhaps the most apt term for this, and indeed many of Leonardo's aeronautical designs, is what one Italian has called a 'sogno di fantasia', a 'fantastic dream'.

Type D.   Powered Ornithopter (1495-97)

Before Leonardo arrives at his last proper aircraft, which exhibits advanced characteristics, there is his only powered machine to consider. It can be dated to the years 1495–97 (Fig. 13), and indeed is only a pipedream of an aircraft; but it is an interesting dream. It shows him, in his middle forties, turning to the only source of power then available to him, the bow-string motor. This, of course, can be made very powerful, but it is of exceptionally short duration. The machine seems to be a standing-type ornithopter, but with only two wings. The motor, roughly indicated in the general sketch, is shown alongside, where the bow mechanism is presented in more detail, with the bow-strings operating through a gear-train. The whole conception of this machine was, or course, charmingly outrageous, and the idea of the pilot periodically having to wind up the mechanism *en route*, so to speak, presents an intriguing vision of desperation! But this design at least shows that Leonardo had been pondering the problem of the ornithopter, and was now thinking in terms of mechanical, rather than muscular, power.

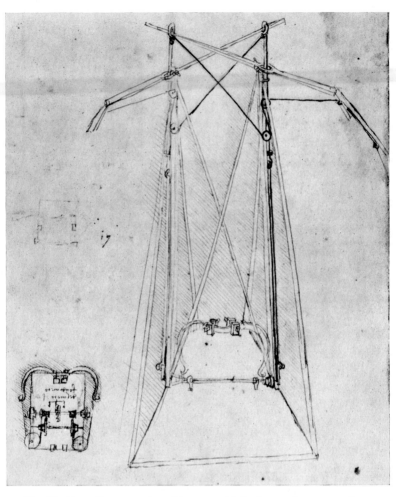

Fig. 13. Type D, powered ornithopter, with its bow-string motor shown separately: 1495–97.

18

Type E. Semi-ornithopters (1497-1500)

It seems to have been after the period of the standing ornithopter—but dates can only be approximate—that the worst of Leonardo's fever of fantasy had burnt itself out; thereafter we find some of the most remarkable of his creative thinking in aviation, as opposed to his obsessional thinking. In what may be called here a semi-ornithopter, we have a prevision of Otto Lilienthal's hang-gliders of 1891–96, where the pilot hung vertically in the centre-section of his machine (Fig. 14). But Leonardo did not here intend his pilot to swing his torso and legs to shift the centre of gravity, and hence exercise a limited control of it in pitch, yaw and roll; nor did he intend to make fixed-wing glides. But there is an interesting thumb-nail sketch and description made later, in 1505 (see Fig. 20), which shows that Leonardo at least became aware of the problem of body-control at that date.

However, the most significant feature in this type of aircraft is the abandonment of the pure ornithopter, in which the whole of each wing is flapped, and the adoption of a partially fixed-wing configuration in which only the outer portion of the wing is flapped. This was another pre-vision of Lilienthal, in this case of his powered gliders of 1895, which incorporated multiple ornithoptering wing-tip slats, in imitation of a bird's primary feathers. In Leonardo's case, it was a question of about half the wing-area being made to flap. This shows, I believe, that Leonardo had been thinking of the structure of a bird's wing, and had realised that the inner part moves far slower than the outer, and therefore produces mainly lift rather than thrust: so he wisely economises on the amount of wing-surface that has to be flapped in relation to the muscular resources of the pilot, and thus concentrates the movement and effort where it can best be utilised.

In the rapid generalised sketches included in Fig. 14, Leonardo does not detail the complicated backward-flapping technique he believed was essential, nor does he show how he would solve the problem of the power transmission system. He simply sketches the general configuration of the machine, and indicates the support of the pilot by having him hang in a sling beneath his crotch. The ornithoptering wing-ends are only capable of bearing downwards. In another drawing (not reproduced here) there is a hint of a solution, where Leonardo has incorporated an oblique hinge, which would flap the end downwards and backwards: he is here very near the true wing movements of a bird, but he does not make the necessary imaginative leap and grasp the propeller function of the outer parts of the wings. But the basic conception of this type of aircraft is advanced far beyond any of Leonardo's former machines.

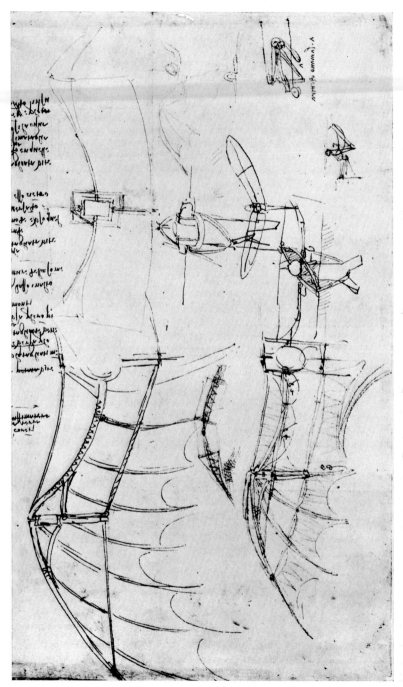

Fig. 14. Type E, semi-ornithopter, with fixed inner wings, ornithoptering outer portions, and 'hang-glider' pilot-position: 1497–1500.

Type F. Falling-Leaf Glider (1510-15)

It is at the very end of his life that we find one of the most intriguing of Leonardo's ideas: it scarcely merits inclusion among his aircraft types; but, in essence, it is an embryo type on its own, and the sole representative in Leonardo's life of the pure gliding principle, however primitive. So I have designated it Type F, the falling-leaf glider, which was clearly inspired by the observation of nature (Fig. 15). Beside the first set of these little sketches—the 'aircraft' consisting of a piece of 'slightly curved' paper—Leonardo writes: "Of the things that fall in the air from the same height, that will produce less resistance which descends by the longer route: it follows that that which

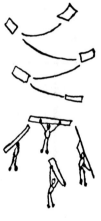

Fig. 15. Type F, falling-leaf glider, showing a series of descent positions: 1510–15.

descends by the shorter route will produce more resistance . . . [the paper] although in itself of uniform thickness and weight, being in a slanting position, has a front that has more weight than any other part of its breadth equal to the front, which can serve as its face; and for this reason the front will become the guide of this descent."

Then, referring to the lower series of sketches, he writes: "this [man] will move on the right side if he bends the right arm and extends the left arm; and he will then move from right to left by changing the position of the arms."

Here are the true beginnings of controlled glider flight, written down and illustrated less than ten years before the death of this remarkable man. One can only regret that such a sequence of events did not occur much earlier in his life, as it could easily have done; and that Leonardo did not initiate the fixed-wing glider proper.

21

Flap-Valve Wings (1486-90)

Leonardo had observed the way in which a bird's feathers overlap to give maximum strength, and to present an air-tight surface on the down-stroke; he concluded that, on the up-stroke, this overlapping structure would, so to say, lift apart, and allow the air to pass easily between the feathers to reduce resistance. In fact, this sequence of events does not take place in bird-flight, and the wing on its up-stroke does not let the air through. But Leonardo's belief led him to invent a highly ingenious flap-like device,* which was to remain aeronautically unexploited until Jacob Degen arrived at the same conclusion during the first decade of the nineteenth century, and incorporated it in his ornithopter. It consisted of stretching a network between the wing-ribs *above* the fabric covering, so that the wing would beat 'solid' on the down-stroke. On the up-stroke the fabric would flap loosely downwards, and un-cover the netting, thus allowing the air to pour through the wing, with only minimal resistance. A general view of a wing with one of the panels uncovered to reveal the netting is shown in Fig. 16; and two detailed drawings of the flap-valves are given in Fig. 17.

---

*The full engineering term for this today is a 'swing flap check valve'.

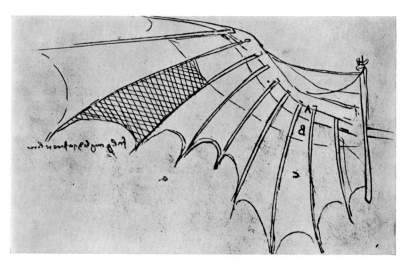

Fig. 16. Ornithopter wing, with one panel stripped to show netting of the flap-valve system: 1486–90.

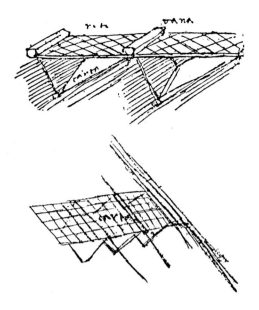

Fig. 17. Two variations of the flap-valve system for ornithopters: 1487–90.

Leonardo was as much fascinated by bat-flight as by bird-flight, as will be seen in these flap-valve wings. His wing-forms in general show evidence of the influence of both. At times he specifically recommends the bat: "remember that your bird [*i.e.* aircraft] should have no other model than the bat, because its membranes serve as an armour, or rather as a means of binding together the pieces of its armour, that is the framework of the wings." But he is just as apt to return to bird construction at other times.

Articulated Wings (1495-97)

Leonardo, from the first, arranged for his wings to be articulated in order to produce—*via* the twisting of the spar—the downward and backward beat. As his ingenuity became increasingly to possess him, he indulged in ever more complex designs for the articulated 'fingers'; his drawings are often beautifully executed, and one of the best—dated 1495-97—is shown in Fig. 18, where Leonardo's twisting and pulley arrangements are admirably set out. But he was soon to abandon these obsessive ingenuities, and progress to simpler and more direct methods.

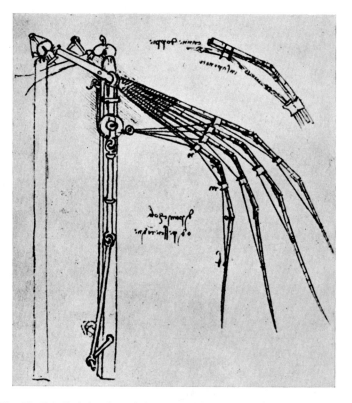

Fig. 18. Detailed drawing of the mechanism for flapping and twisting the wing of an ornithopter: 1495-97.

Flight-Control (1499-1505)

It has already been shown that Leonardo, during the period 1486–90, designed the first control system of history, when he incorporated a cruciform tail-unit in one of his prone ornithopters; this was to be operated by the pilot moving his head: the end of the pole which worked the tail-unit being secured to his head by a bandage. Now, at the turn of the century, we find a single small sketch of a horizontal tail-surface to be used as both elevator and rudder: it is dated about 1499–1503 (Fig. 19). But Leonardo's ingenuity has in this case run away with him; instead of the comparatively simple operation of raising or lowering the horizontal rear surface to act as the elevator, he has here proposed a device whereby movements of the pilot's head would produce the elevator effect by increasing or reducing the area of the surface, and the rudder effects by leaning his head to one side or the other. He writes beside the

Fig. 19. Diagram of a horizontal tail-unit, for spreading, contracting or swivelling the surface: 1499–1503.

drawing: "Here the head *n* is the mover of this helm; that is, that when *n* goes towards *b*, the helm becomes widened; and when it goes in the opposite direction, the tail is contracted; and similarly when *f* is lowered, the tail is lowered on this side; and so lowering itself on the opposite side it will do the same." The mechanism is certainly ingenious: the convergent ribs at the front end of the device are fixed to a swivelling hinge, and the pilot's head-band has a ring on each side moving along each rib of the elevator. The idea comes direct from the intensive study of bird flight that Leonardo was then making, and imitates the bird's ability to enlarge or contract its tail, fan-wise, and to swivel it. But Leonardo has gone against the instinctive movements which a human pilot would make; that is to say, he would duck his head to dive, and pull it back to climb—the opposite to what Leonardo suggests here. His cruciform tail-unit of 1486–90 was a far better arrangement; now, in his desire to imitate the birds, he has moved away from practicality, as others were to do later on.

But in his complete illustrated 'booklet' on bird flight, *Sul Volo degli Uccelli*, written in 1505, Leonardo has for a moment come very close to anticipating Lilienthal's hang-glider technique of flight-control of 1891–96; for, on folio 6 (5) recto, he has made a rapid sketch (Fig. 20) of a man standing in an ornithopter, with these accompanying remarks: 'The man in a flying machine [has] to be free from the waist upward in order to be able to balance himself as he does in a boat, so that his centre of gravity and that of his machine may oscillate and change where necessity requires, through a change in the centre of its resistance." This realisation of the relationship between the centre of

Fig. 20. Sketch diagram of an ornithopter pilot effecting control by body-movements: 1505.

gravity and the centre of pressure is typical of the growing clarity of conception—where aviation is concerned—that Leonardo is achieving in the last years of his life. He is here anticipating the flight-control technique of Lilienthal, in 1891–96, in which the pilot throws his hanging torso and legs in the direction in which he wishes to travel, thus shifting the centre of gravity, and with it the centre of pressure. Leonardo here adopts the unwise technique of exerting this shift from above, and hence making for a de-stabilising situation: but, if he had lived, he would almost certainly have realised this, and adopted the hang-glider technique of flight-control prior to the achievement of movable control surfaces, which he had already envisaged with his cruciform tail-unit, but unfortunately abandoned.

Wing-Testing Rigs (1485-90)

Leonardo has left a few sketches of devices for testing and measuring the forces acting on his ornithoptering wings, the best known being the vigorous drawing reproduced in Fig. 22. But he has left a slightly earlier and more interesting study (Fig. 21), where he invisages a man with a wing placed on one pan of a pair of scales, and weights on the other. He writes beside this sketch: "And if you wish to ascertain what weight will support this wing, place yourself upon one side of a pair of balances, and on the other place a

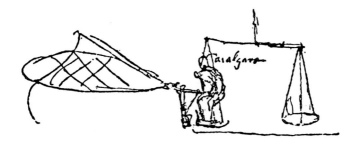

Fig. 21. Wing-testing rig on scales, for an ornithopter wing: *c.* 1485.

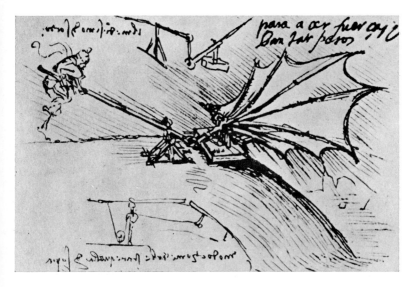

Fig. 22. Wing-testing rig for an ornithopter wing: 1486–90.

corresponding weight, so that the two scales are level in the air; then if you
fasten yourself to the lever where the wing is, and cut the rope which keeps it
up, you will see it suddenly fall; and if it required two units of time to fall of
itself, you will cause it to fall in one by taking hold of the lever with your
hands; and you lend so much weight to the opposite arm of the balance that
the two become equal in respect of that force; and whatever is the weight of
the other balance, so much will support the wing as it flies; and so much the

more as it presses the air more vigorously.'' These tests would be somewhat misleading, since the pilot of an ornithopter has to flap two wings simultaneously, and the expenditure of a man's energy in beating down two wings would be much in excess of double the force necessary to beat down one, owing to the metabolic and energy-replacement problems involved.

The more elaborate sketch (Fig. 22) is only to show the transmission system, and one feels Leonardo has been charmingly carried away by his sense of the dramatic.

Type G. Finned Projectiles (c. 1485)

The earliest device designed by Leonardo which involved aerodynamic considerations was probably the finned projectile, of which four notable examples are to be found in the manuscripts; these were drawn about 1485, when he was in the service of Ludovico Sforza. The three most sophisticated are shown in Fig. 23: each is equipped with four directionally stabilising fins at the rear. From a description Leonardo has written beside the third example—which is more dart-shaped—he intended to launch these projectiles from a ballista, and the fins would undoubtedly increase their range and accuracy. The example shown in (a) would have probably suffered aerodynamically from the presence of the two knobbed horns which were to act as impact fuses: the left-hand one (b) would be the most directionally stable, with its streamlined nose, and the fins faired neatly into the shell: the right-hand example (c) would probably have suffered from turbulence set up by the cut-away shell at the point of the fin attachment. There is no way of telling if any of these designs were actually built and tested. Such astonishingly modern-looking weapons were not to be rivalled until Cayley came to design and test his finned projectiles in 1804–05.

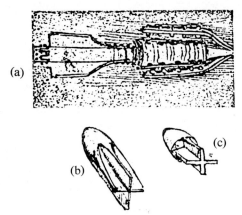

(a)

(b)

(c)

Fig. 23. Type G, finned projectiles; three weapons with various fin arrangements: c. 1485.

Type H.   Parachute (1483-86)

About the same time (1483–86) came Leonardo's design for a parachute (Fig. 24): this is the world's first example, and is of pyramidal configuration, and Leonardo wrote beside it: "If a man have a tent made of linen, of which the apertures have all been stopped up, and it be twelve braccia across and twelve in depth, he will be able to throw himself down from any great height without sustaining any injury". An interesting feature of the design is the pole running down from the apex of the canopy: Leonardo was concerned here with making his overall structure rigid, with the shroud lines tied to the end of the pole. As the man hangs by his arms beneath the jointure, he would not be so badly affected by the oscillations from which this type of parachute would suffer.

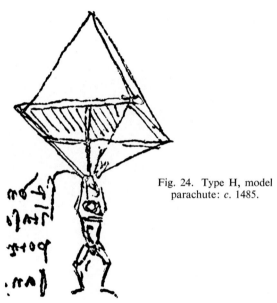

Fig. 24. Type H, model parachute: *c.* 1485.

It would be interesting to know whether Leonardo ever had in mind the escape from aircraft, as well as the more general idea he describes*; and also why he did not employ a variation on the theme of the parasol, rather than the tent. As Leonardo's sketch remained unknown until late in the nineteenth century, it played no part in the development of modern parachutes, which are all derived from the parasol, the first air-to-ground jump being made by Garnerin in 1797.

---

*He recommended flying over water, which he seems to have felt would be soft to land upon, and suggested skins and bladders for flotation.

Type I. Helicopter Model (1486-90)

Leonardo's admirable little helicopter design (Fig. 25), of which he seems to have made a successful model, could until recently claim to be the first helicopter of history. But it has now become known that the model helicopter was understood and successfully made before Leonardo's machine, in the form of a toy whose rotor was based on windmill sails, and which was sent spinning up into the air by pulling sharply on a cord wound tightly round the shaft; the earliest known toy of this kind dates from about 1320–25.

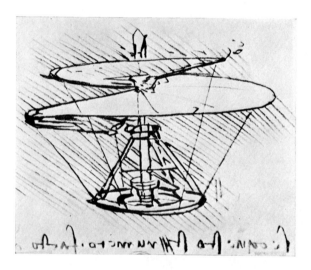

Fig. 25. Type I, helical screw helicopter model: 1486–90.

Leonardo's machine carried its power-unit on board, but was made in the unsatisfactory form of a helical screw, instead of rotor blades. This design, as with the rest of Leonardo's work, did not become known until long after the model helicopter, powered by a variety of prime movers, was firmly established on the aeronautical scene. But if the report is true—and one has no good reason to doubt it—Leonardo played an indirectly vital part in the development of the modern helicopter; for it is said that it was an illustration of this model that was shown to Igor Sikorsky by his mother, and that the sight of it inspired him later to take up the study of the helicopter, in which he has led the world.

Leonardo never returned to this talented conception; and the sketch is rapidly thrown off on one of his pages, with the comment: "I find that if this instrument made with a screw be well made—that is to say, made of linen of which the pores are stopped up with starch—and be turned swiftly, the said screw will make its spiral in the air and it will rise high. Take the example of a wide and thin ruler whirled very rapidly in the air, and you will see that your arm will be guided by the line of the edge of the said flat surface. . . . You can make a small model of pasteboard, of which the axis is formed of fine steel wire, bent by force, and as it is released it will turn the screw".

The sentence before the last is one of the most tantalising in Leonardo's writings, for here is the germ of the modern active airscrew. It is also strange—we being now wise in hindsight—that he never seems to have thought of applying the Archimedean helicopter screw to work horizontally as an airscrew for a flying machine; but ornithopters ruled his mind.

This helicopter model—if we assume it was made and flown, as I think we may—has the curious distinction of being the world's first powered aircraft; that is to say, the first aircraft to become airborne with its power-unit incorporated in its structure.

Inclinometer for Aircraft (c. 1485)

Early in the period of his aeronautical speculations, and designs for aircraft, there is a surprising little drawing (Fig. 26) for an inclinometer specifically intended for aircraft: it is an instrument which indicates the attitude of a machine relative to the horizontal, and shows Leonardo considering what may even be felt as comparatively academic matters, considering the "state of the art" of flying at that time. The device consists of a small pendulum hanging in a frame which was to be fixed to the aircraft.

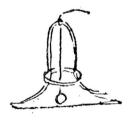

Fig. 26. Design for an aircraft inclinometer: *c*. 1485.

Bird-Flight

Leonardo seems to have carried out his most intensive investigations into bird flight when he returned to Florence after the French invaded Milan in 1499. In 1505 he wrote a small complete treatise on the subject—*Sul Volo degli Uccelli* (On the Flight of Birds)—illustrating it with rapid and vivid little sketches (Fig. 27). There is no space here to make more than a brief comment on his work in bird flight: he is often penetrating and shrewd in his descriptions of birds in their various kinds of flight, including gliding, soaring and flapping, as well as manoeuvre, along with the properties of the air: he also

Fig. 27. Illustration from the treatise on bird flight (*Sul Volo degli Uccelli*): 1505.

includes discussions on the relationship between the centre of gravity and the centre of pressure. But he was—as remarked before—unaware of the technique by which birds propel themselves through the air; and it is even more surprising that in this later period of deep study and contemplation, Leonardo

Fig. 28. Apparatus for determining the centre of gravity of a bird: 1505.

did not look at the emarginated feathers of some of the birds he dissected, and wonder why this particular structure was provided by nature; and why he did not envisage the twisting airscrew effect that would be produced by the down-beat of these emarginated feathers, with so much vane surface aft of their shafts, and so little forward. But his brilliant mind had encompassed so much, that there were inevitably blind spots along the way. As a typical example of his careful experimental approach to many aspects of bird-flight, we may illustrate his device for determining the centre of gravity of a bird (Fig. 28).

Airscrew-operated Smoke-Jack (1480-82)

One of the most interesting devices of a semi-aeronautical nature to be found in Leonardo's manuscripts, is his drawing of an airscrew-operated smoke-jack to turn a roasting spit (Fig. 29). He may not have invented this device, but he sets out here a neat piece of machinery which shows that he understood the passive action of the airscrew early in his aeronautical inventive period. It raises again the problem of why Leonardo did not experiment further with the air or water screw; but they did not seem to interest him very much; and even the water-turbine he once sketched seemed only of passing interest.

33

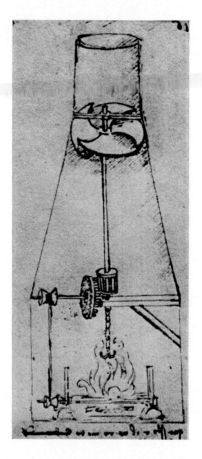

Fig. 29. Design for an airscrew-operated smoke-jack to work a roasting spit: 1480–82.

Did Leonardo attempt a Flight?

The answer to this question must be an unequivocal 'no'. It is just possible, but highly unlikely, that Leonardo built one of his prone ornithopters, and lay in it fruitlessly flapping its wings: its weight alone would have kept it firmly earthbound. But he certainly never had himself launched into the air; the machine's weight, together with the rapid movement of the centre of pressure rearwards the moment he made the first down-and-back-beat of the wings, would have immediately led to a disastrous crash, with death or serious injury for the pilot.

There is a hazy tradition which came third hand from Giralamo Cardano that Leonardo attempted a flight ("Leonardo da Vinci also tried to fly, but he, too, failed"). But the curious passage in one of the note-books implying that a flying machine was constructed refers, I think, to a model, as said, and not to a full-size machine, as sometimes believed: "Close up with boards the room above, and make the model large and high, and you will have space upon the roof above . . . if you stand upon the roof at the side of the tower, the men at work upon the cupola [*i.e.* of the Cathedral] will not see you". A much more mysterious note reads, briefly: "To-morrow morning, on the second day of January 1496, I will make the thong (soatta) and the attempt"; but no one knows to what he refers, whether it be a flying machine, a model, or some undertaking quite apart from aviation.

There remains the possibility that Leonardo built a full-size mock-up of a machine to see how it would appear, and to provide him with data about positioning the pilot, and the transmission system. But even this was not likely, as Leonardo had an almost pathological inability to complete any undertaking, as seen in his work on the Sforza monument and other projects.

Leonardo and the Hot-air Balloon

Leonardo has also been credited by some enthusiasts with the invention of the hot air balloon. The whole claim is based on a story in Vasari's *Lives of the Painters*, etc. (1550), which tells of Leonardo's visit to Rome for the coronation of Pope Leo X. Vasari, obviously basing his description on distant hearsay, says that Leonardo "made a paste of wax, and, while he was walking, shaped very thin hollow animals, blowing into which, he made them fly through the air. But when the wind ceased, they fell to the ground." First of all, the human breath is not hot enough to raise anything: secondly, even if a little fire had been hung beneath the 'animals', the hot air so generated could not have power enough to lift such objects, no matter how thin the wax, to say nothing of the wax melting; and thirdly, even if the creatures had been made out of paper, and a small fire (candle or wick) been placed underneath, the hot air would still have been powerless, because at that time no paper was made thin and light enough to be raised unless the 'balloon' had been of giant proportions, according to the cube square law. Only the thinnest 'tissue paper' could have produced the result Vasari describes, and even then with a fire, and not human breath. What Leonardo probably did was to cut out and fly little kites in the form of animals, which would of course fall down when the wind ceased, unless they were towed.

The Manuscripts

Leonardo's manuscripts, mostly in the form of note-books, are all written with the left hand in reverse, from right to left, as if in a mirror. No agreement has been reached about the reason for this. We can quickly dismiss the idea that he wrote in such a way to preserve secrecy, as the veriest simpleton would realise he could hold a mirror in front of the pages and read them straight off. There is some reason to believe that early in his life Leonardo had injured his right hand; and medical evidence suggests that if a right-handed man has to learn to write left-handed, his instinct—unless forcibly corrected—is to write backwards across the page, from right to left.

The reader may have noticed that some of the drawings I have reproduced appear to be the 'right way round', and others Leonardo's 'wrong way'. The reason for this is that the choice of facsimiles I have used depended on their clarity of line; some were printed one way by the former editors, some the other. I have chosen the clearest in each case.

When Leonardo died on May 2nd, 1519, at the small castle of Cloux, near Amboise in France, the faithful companion and friend who was with him was Francesco da Melzi. Shortly before his death, Leonardo made a will in which he left the entire corpus of his manuscripts to Melzi. Never was there a more charming and human gesture; never was there a more ultimately disastrous gesture.

Melzi returned with this fabulous hoard of material to his villa at Vaprio (between Milan and Bergamo): here he guarded his treasure affectionately and jealously, and resisted all offers to part with it, for over fifty years. But history can only pronounce the severest censure upon Melzi, kind and loyal man though he was. For he made no effort to ensure that Leonardo's manuscripts were left in reliable hands after his death. Melzi well knew that his son Orazio had not the slightest interest in the arts and sciences; yet he deliberately left all the manuscripts in his will to this deplorable man. Orazio Melzi "could not have cared less" (in the modern phrase) what happened to this priceless collection of documents, and he allowed various people who appeared on the scene to beg, borrow, and all but steal, anything they wanted. Orazio permitted the dispersal of the Leonardo manuscripts, a dispersal which led to damage, loss, and vandalism, although happily the bulk of the material has survived, thanks to the care of many devoted men over the centuries. Thus at Orazio's door lies the immediate blame; but it is at the door of old Francesco da Melzi that the real blame must lie, for it was he who betrayed the deep trust Leonardo placed in him by allowing this massive evidence of his genius to fall into the hands of an unworthy son.

Leonardo's Influence in Aviation History

It was said at the beginning of this booklet that Leonardo's work in aviation had no influence on the history of aviation, owing to none of it being known to the pioneers until late in the nineteenth century. This regrettable state of affairs was due to Melzi's inaction during the fifty years he possessed the manuscripts, and to his irresponsible action in leaving them to the wretched Orazio. There is evidence that Leonardo wished his work to be published; and if Melzi had convened a meeting of scholars after Leonardo's death, they would most probably have arranged a series of publications. If that did not prove feasible during those fifty years, then the least Melzi could have done would have been to bequeath the material to a university, where it would have become widely known, and probably published.

As matters turned out, it was not until Napoleon looted the Leonardo manuscripts from the Ambrosiana Library at Milan—the *Codice Atlantico*, and what are now known as Manuscripts A to M—that they were properly examined by the great French scholar J. B. Venturi, who published a short resumé of Leonardo's scientific work in 1797, entitled *Essai sur les Ouvrages Physico-Mathématiques de Léonard de Vinci*. Interestingly enough, although Venturi must have studied Leonardo's notes on flight, he did not include any mention of the aeronautical material, possibly because he considered them too visionary.

The first actual publication of any of Leonardo's work in aviation was in 1793, when C. G. Gerli included in his *Disegni di Leonardo da Vinci* a few plates reproducing—re-drawn and engraved—some twenty of Leonardo's aeronautical sketches, including the parachute. The work was intended for artists and scholars, and had a very limited circulation: the flying sketches had only a curiosity value for the few who ever saw them. A new edition was published in 1830, with the number of flying items reduced, and the parachute excluded. The practical parachute—derived from parasol ancestors—had been in use for dropping animals from balloons since 1784, before the first live drop by Garnerin in 1797 already mentioned.

In 1838 G. Libri, in his *Histoire des Sciences mathématiques en Italie*, made a passing reference to Leonardo having studied bird flight with a view to making human flight possible; and in 1872 another Italian author, G. Govi in his *Saggio delle Opere di Leonardo da Vinci*, touched on Leonardo's interest in flying, but dismissed it as an extravaganza; but later (in 1881) it was Govi who first published Leonardo's helicopter design.

By the time contemporary scientists saw this ingenious little helicopter sketch, European inventors had already been helicopter conscious ever since Cayley published (in 1809) his version of the Launoy and Bienvenu helicopter model of 1784. After that, helicopter toys became growingly popular with the pioneers.

It was not until 1874 that the first publication appeared which properly established Leonardo as the earliest man to make scientific investigations into flight. This was an illustrated article by the French aeronautical pioneer

Hureau de Villeneuve, entitled 'Leonardo de Vinci, aviateur', and it appeared appropriately enough in the September issue of the excellent periodical *L'Aéronaute*. It presented, of course, only a cursory treatment of the subject; but it marked the first occasion on which the world was properly introduced to Leonardo's work, and told of his unique standing in the history of aeronautics.

In 1881 there started the monumental task of publishing facsimiles and transcriptions of all Leonardo's manuscripts, a task largely completed by 1914, and shared by scholars from Italy, France and Great Britain.

It was not until 1936* that a full survey of all the aeronautical manuscripts was published in Rome by R. Giacomelli, entitled *Gli Scritti di Leonardo da Vinci sul Volo*; to be followed in 1952 by an equally important work by A. Uccelli, *I Libri del Volo di Leonardo da Vinci*.

---

*Ivor Hart made the first serious summary in English of Leonardo's flying work, and it appeared in the *Journal of the Royal Aeronautical Society* in 1923.

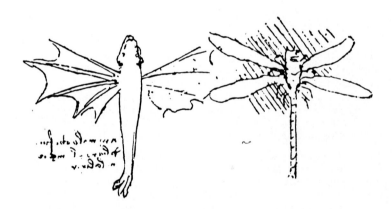

Sources of the Illustrations

Fig. 1.	Royal Library, Turin.
Figs. 2, 3.	Cod. Atl. fol. 276, recto-b.
Fig. 4.	MS.B, fol. 74, verso.
Fig. 5.	Cod. Atl. fol. 302, verso-a.
Fig. 6.	MS.B, fol. 75, recto.
Fig. 7.	MS.B, fol. 79, recto.
Fig. 8.	Cod. Atl. fol. 327, verso-a.
Fig. 9.	Cod. Atl. fol. 313, verso-a.
Fig. 10.	Cod. Atl. fol. 313, recto-a.
Fig. 11.	MS.B, fol. 80, recto.
Fig. 12.	MS.B, fol. 89, recto.
Fig. 13.	Cod. Atl. fol. 314, recto-b.
Fig. 14.	Cod. Atl. fol. 309, verso-a.
Fig. 15.	MS.G, fol. 74, recto.
Fig. 16.	MS.B, fol. 74, recto.
Fig. 17(a).	Cod. Atl. fol. 309, verso-b.
Fig. 17(b).	Cod. Atl. fol. 311, verso-a.
Fig. 18.	Cod. Atl. fol. 308, recto-a.
Fig. 19.	MS.L, fol. 59, recto.
Fig. 20.	Sul Volo, fol. 6(5) recto.
Fig. 21.	Cod. Atl. fol. 381, verso-a.
Fig. 22.	MS.B, fol. 88, verso.
Fig. 23(a).	Windsor, No. 12, 651.
Fig. 23(b).	Cod. Arundel, fol. 54, recto.
Fig. 24.	Cod. Atl. fol. 381, verso-a.
Fig. 25.	MS.B, fol. 83, verso.
Fig. 26.	Cod. Atl. fol. 381, recto-a.
Fig. 27.	Sul Volo, fol. 8(7) recto.
Fig. 28.	Sul Volo, fol. 16(15) verso.
Fig. 29.	Cod. Atl. fol. 5, verso-a.

Cod. Atl.	*Codice Atlantico*; Ambrosiana Library, Milan.
MSS.B,G,L.	Notebooks in the Library of the Institut de France, Paris.
Sul Volo	*Sul Volo degli Uccelli*; Royal Library, Turin.
Windsor	Royal Library, Windsor Castle.
Cod. Arundel	Arundel manuscript No. 263; British Museum.

Further Reading

There is, unfortunately, no modern work in English on Leonardo da Vinci's aeronautics other than the present booklet. The two comprehensive Italian works—which, incidentally, do not classify his work in the way attempted here—are noted on page 38. But Leonardo should always be viewed in his totality, as a complete and highly complex character; as such, he is very well served in the few books listed below.

CLARK, SIR KENNETH. *Leonardo da Vinci.* 2nd ed. Cambridge, 1952. (Republished as a Pelican book).
  The classic biography, and interpretation of Leonardo; it is primarily directed towards his artistic work, but is outstanding for its treatment of Leonardo's character and personality in all its complexity.

RICHTER, J. P. *The Literary Works of Leonardo da Vinci.* Compiled and edited by J. P. R. [the original Italian is given opposite each translation]. 2nd ed. 2 vols. London, 1939.
  An excellent translation, but it includes very little on the subject of flight.

MACCURDY, E. *The Notebooks of Leonardo da Vinci.* Edited and translated by E. MacC. 2 vols. London, 1948.
  The fullest selection of textual excerpts from the manuscripts, including the passages on flight.

HART, IVOR B. *The World of Leonardo da Vinci: Man of Science, Engineer, and Dreamer of Flight.* London, 1961.
  The best account of Leonardo's work in science and technology, with 125 illustrations.

HART, CLIVE. *The Dream of Flight.* London, 1972.
  This work includes a chapter on Leonardo's aeronautics, particularly valuable regarding his work on bird flight.

LEONARDO DA VINCI. 2 vols. London, 1964.
  This huge book, published in Italy, Germany, and Britain, provides the largest number of illustrations (some 2,000) of all aspects of his artistic and scientific work: it is also the only book to bring together the illustrations according to their subject matter, so that one finds groups of illustrations—with text—on his aeronautics, anatomy, architecture, etc.

FRIEDENTHAL, R. *Leonardo da Vinci: a pictorial Biography.* London, 1959.
  A short introduction to Leonardo and his work, with 125 illustrations, but very poor on his aeronautics (the helicopter is called successively an "airscrew" and an "air gyroscope").

Grateful acknowledgement is accorded to Messrs Jonathan Cape for their permission to quote from MacCurdy's translations on the subject of flight.